Cultivated Cuisine:

The Future of Lab-Grown Meat

by Gordon Rayner

Formatted, Converted, and Distributed by eBookIt.com
http://www.eBookIt.com

ISBN-13: 978-1-4566-4106-1 (paperback)
ISBN-13: 978-1-4566-4105-4 (ebook)
ISBN-13: 978-1-4566-4107-8 (audiobook)

Dear Esteemed Reader,

Thank you immensely for choosing this book to join your collection. We imagine that you've already embarked on an exploration of ideas within these pages, and we couldn't be happier about it!

Now, if you find yourself chuckling, pondering, or even debating with the words in front of you, we'd absolutely love to hear about it. If you can spare a few moments to pen down your thoughts in a review, we would be as delighted as a dictionary on a spelling bee!

An Amazon review would be excellent - but hey, we're far from picky. Whether it's a scribble on the back of a grocery list, a tweet, or even a message in a bottle (though that might take a while to reach us), your feedback is gold.

Writing a review might not be as fun as a spontaneous dance-off, but we promise it'll bring grins to our faces, warmth to our hearts, and incredibly valuable insights to future readers.

With Gratitude,

Bo Bennett, PhD
Publisher
Archieboy Holdings, LLC.

Table of Contents

Introduction

Welcome to an exploration of culinary innovation that takes a deeper dive than any sous-vide machine has gone before. You're about to embark on a journey where the briny seas of traditional food production part to reveal an undercurrent of change so compelling, it might just redefine our dinner plates. Today, we're staring down the microscope at the brave new world of lab-grown meat, or, as it prefers to be known after a glass or two of red, cultured meat.

In the grand scheme of things, this meaty revolution is akin to that awkward phase between tossing raw mammoth steaks onto the fire and discovering the joys of medium-rare. You see, lab-grown meat isn't a sci-fi writer's midnight snack — it's as real as the pickle in your hamburger. But it doesn't moo, cluck, or oink. Instead, it grows, quite peacefully, in a bioreactor, oblivious to the machinations of the outside world, like a teenager on a particularly gripping video game binge.

Now, before you furrow your brows and consider bolting, let me assure you: we're not in mad scientist territory here. This is a tale of innovative minds peering into a petri dish and seeing a world where our hunger for meat no longer exacts such a heavy toll on the planet or its creatures. This is no mere Frankenfood story, but a quest for something altogether more meaningful: a way to secure our carnivorous cravings while ensuring a sustainable future.

Just as alchemists once sought to transform base metals into gold, our modern culinary magicians are turning humble cells into succulent steaks. It's a whole new take on having your cake and eating it too, and while it might not please Goldilocks just yet, the promise it holds is simply too tantalizing to ignore.

In the pages that follow, we'll journey together through the intricate landscape of this food frontier, gazing in awe at the science that transforms a smattering of cells into a gourmet meal. We'll peek behind the lab doors at the strides in biotechnology that make this all possible, akin to a backstage pass at a rock concert, but with fewer guitars and more petri dishes.

But, like any tale worth its salt, this one doesn't lack for conflict. This burgeoning field grapples with its share of challenges, from technical obstacles to ethical quandaries that would leave even the most steadfast philosopher scratching their head. And yet, much like the little train that could, it chugs along, driven by the promise of what might be.

We'll also delve into the world of public perception, where lab-grown meat dances a delicate minuet with public opinion. Like a new kid on the block trying to fit into an old, tight-knit neighborhood, cultured meat has its work cut out for it. The question on everyone's lips (aside from 'How does it taste?') is: How does this new arrival fit into our established food culture?

And then there's the ripple effect. This isn't just about creating a new type of food. No, it's about the cascading impacts that could reshape agriculture, industry, and even our global economy. Like dropping a pebble into a still pond, the ripples of cultured meat will spread far and wide, disrupting the status quo in ways we can only begin to fathom.

So strap on your lab goggles and fasten your aprons, because we're about to embark on a journey that bridges the gap between science and supper. The future of food is here, and it's cultured, not cultivated. Ready for the first bite?

Chapter 1:
A New Age of Meat

Our culinary journey begins where it must: the dawning of a new epoch in the meaty annals of human history. And no, we're not talking about slapping the next 'impossible' label on a plant-based substitute or conjuring up another vegan 'miracle'. We've embarked on a rather more ambitious endeavor – it's nothing short of harnessing the cellular building blocks of life to create actual, factual, bona fide meat, minus the onerous business of livestock rearing. It's akin to wanting a family but bypassing the trials of raising an unpredictable toddler, instead heading straight for the well-adjusted adult phase. So, dust off your finest dinner attire and brace yourself for the grand reveal of the guest of honor at our culinary party – lab-grown meat.

What is Lab-Grown Meat?

If one were to think of lab-grown meat as a fancy guest at a food party, there would surely be whispers around the buffet table. Is it some strange alien dish, a Frankensteinian creation, or just another science fiction story come to life? Rest assured, it's not quite as ominous as it sounds, nor does it involve any extraterrestrial barbecue techniques or monstrous cooking methods.

Lab-grown meat, scientifically termed 'cultured meat', has its roots in the same place as that juicy sirloin or succulent chicken breast - the cell. Indeed, this is meat grown from cells, typically muscle cells, that are taken from an animal and nurtured into a form we recognize as food. No science fiction, just good old-fashioned biology with a modern twist.

The technique behind it is a bit like growing a plant from a cutting, if that plant was actually a steak and the gardener wore a lab coat. It begins with a small sample of animal cells, which are then fed a nourishing concoction of nutrients, vitamins, and sugars, sort of like an energy drink for cells, encouraging them to grow and multiply.

These cells go about their business in a cozy environment called a bioreactor, a sophisticated device that emulates the conditions inside an animal's body. Picture it as a high-tech petri dish meets a balmy cruise vacation, where cells can kick back, relax, and multiply to their heart's content.

Now, in an ideal world, these pampered cells would form into muscle tissue, layering together in a three-dimensional network of fibrous meaty goodness. Of course, just like a good soufflé, the process requires some delicate handling, precision, and a dash of culinary magic. But instead of a warm oven and crossed fingers, this process utilizes bioengineering techniques and the natural propensity of cells to form tissues.

What's particularly fascinating about this process is that it's all done without the usual trappings of animal farming – no fields, no barns, and most importantly, no need for the animal to be, shall we say, 'eternally indisposed'. This is the essence of lab-grown meat: real, authentic meat, produced without the associated environmental and ethical concerns of traditional livestock farming.

But how does it taste, you ask? Ah, the million-dollar question, and quite rightfully so. After all, we're not just aiming to replicate the protein content of meat. We want that juicy, tender, succulent experience that sets our taste buds singing the 'Hallelujah Chorus'. And the answer, dear epicure, is that lab-grown meat does indeed hold its own at the gastronomical rodeo. It looks, cooks, and tastes like meat because, in essence, it is meat, just with a dash of scientific flair.

So, there you have it - a whirlwind tour of lab-grown meat, or cultured meat if we're being formal. It's a brave new world where biology meets gastronomy, under the watchful eyes of lab coats instead of chef aprons. Here's to the future, one cell at a time.

The History of Cellular Agriculture

Let's take a stroll down memory lane, where the whiteboard markers are as fresh as a summer's dawn and the echo of visionary ideas still reverberate in the halls of academia. This is where our story of cellular agriculture begins, with the marriage of science and gastronomy in a way that would make even the most unlikely rom-com proud.

As far back as 1931, Winston Churchill, in between bouts of leading a nation and penning an occasional book, casually predicted that we'd all be growing chicken breasts and wings separately in a span of 50 years. His timetable was a tad optimistic, but his vision was not far off. The idea of producing meat without raising animals was considered a bold and tantalizing prospect, akin to flying cars and colonizing Mars.

Fast forward a few decades and enter Jason Matheny, a man for whom the 'business-as-usual' of animal farming was less palatable than an undercooked hamburger. His research into cellular agriculture at the University of Maryland in the early 2000s paved the way for the burgeoning field that we see today. He saw the potential for growing meat in a lab long before it was fashionable or technologically feasible, a testament to his vision or his distaste for conventional farming methods, or perhaps a bit of both.

Meanwhile, a team of Dutch scientists, led by Mark Post at Maastricht University, were cooking up something special. Their goal? To create the world's first lab-grown burger, a quest worthy of a culinary Indiana Jones. This scientific saga culminated in 2013 when they unveiled their masterpiece - a burger made entirely from cultured bovine cells. The price tag was as hefty as the achievement itself, ringing in at a cool $325,000. Not exactly value menu material, but then again, we're talking about a scientific marvel here, not a 2 am drive-thru order.

The Maastricht triumph sparked a global interest in cultured meat, prompting a surge in research and development and seeding numerous startups. It was the 'Eureka!' moment that signaled the realm of possibility, akin to the Wright brothers at Kitty Hawk, proclaiming that, indeed, we can take to the skies.

The progress since that expensive burger has been nothing short of astounding, with leaps and bounds made in refining the process, reducing costs, and improving the quality of lab-grown meat. Scientists today can choose from various animal cells and even manipulate the taste, texture, and nutritional content of the final product. It's a bit like painting with a biological palette, allowing for a creativity that would make Picasso salivate.

But cellular agriculture isn't limited to meat alone. Over the past decade, researchers have extended their science-y tendrils into other animal products like milk, eggs, and even leather. These efforts have birthed an entire industry devoted to developing animal-free versions of these products, with the hope of reducing our reliance on traditional animal farming.

Despite this dazzling progress, the road to commercial viability for lab-grown meat and other cultured animal products has been a slow simmer rather than a rapid boil. The field still faces significant challenges, such as scaling up production, improving cost efficiency, and convincing a skeptical public to accept these new products. It's like trying to persuade your grandma to switch from her landline to a smartphone - daunting but not impossible.

The history of cellular agriculture is a tale of scientific innovation, perseverance, and, dare we say, a dash of audacity. It's a journey that has the potential to revolutionize our food system, combat climate change, and improve animal welfare. So, as we saunter further into this culinary adventure, it's worth savoring how far we've come from Churchill's early musings and looking ahead to the delicious possibilities of the future.

Chapter 2:
The Science Behind Cultured Meat

And so, we venture into the heart of the matter: the science of cultured meat. If you've ever wondered how one goes from a humble cell to a sizzling steak, then you're in the right place. It's a journey that could baffle even the most tenured professor, or make a biophysicist feel like they're deciphering an alien language—albeit, a language with the delectable, mouth-watering vocabulary of ground beef, filet mignon, and barbecue ribs. In this chapter, we're going to take a close (and I promise, not too microscopic) look at the wonders of cellular biology, the high-tech wizardry that transforms biopsies into burgers, and the innovations in biotechnology that are helping to bring this remarkable science to your dinner plate. By the end, not only will you understand the cellular waltz that takes place within your future burgers, but also why some scientists might choose a Petri dish over a barbecue grill. Buckle up and remember, lab coats are optional; an appetite for knowledge, mandatory.

Cellular Biology 101

Before we delve into the meat of the matter, it's important to ground ourselves with a solid understanding of cellular biology, or as some prefer to call it, "the science of the incredibly tiny". As much as we'd like to think that our mastery over fire and the invention of the cheeseburger sets us apart, humans, like all living organisms, are fundamentally just an elaborate symphony of cells. So, let's roll up our metaphorical lab coat sleeves, squint at the proverbial microscope, and dive in.

Imagine, if you will, a bustling city filled with factories, transport systems, power plants, recycling centers, and countless citizens each performing their own specialized jobs. Now, shrink that city down to a microscopic scale, and you have a cell. Cells are the fundamental units of life, the smallest structures capable of performing all the activities necessary for life. From the neurons that help you ponder the meaning of life (or simply what's for dinner) to the muscle cells contracting as you reach for that tantalizingly placed jar of cookies, cells make it all possible.

The star of our show, however, is a special kind of cell called a stem cell. These are the wild cards in the pack, the jacks-of-all-trades. They have this fantastic party trick called "differentiation". Simply put, these cellular chameleons have the potential to become any type of cell they please, be it a heart cell, a kidney cell, or yes, a meat cell. They are the little engines of potential that are going to power our journey from biopsy to burger.

Now, every city needs a rulebook, a set of instructions that keeps the urban machinery running smoothly. In a cell, this role is performed by deoxyribonucleic acid, or as it is more commonly known (and mercifully so), DNA. It's a bit like a cookbook, holding all the recipes that a cell needs to function. For our cultivated meat, this DNA cookbook will hold the instructions for building a delicious, juicy muscle cell perfect for grilling.

The cell's factories, where these DNA instructions are turned into tangible products, are called ribosomes. They read the recipe, gather the ingredients, and whip up everything a cell needs to function, from the enzymes that perform chemical reactions to the structural proteins that give a cell its shape. In our cultured meat scenario, these ribosomes will be the mini-chefs cooking up our meat cells.

Speaking of structure, let's chat about extracellular matrix, the cell's equivalent of scaffolding. It's not just a lifeless structure, it's a dynamic environment that helps guide and influence the behavior of the cells. You could think of it as the world's tiniest construction site, where cells are both the builders and the bricks, guided by this matrix. For our cultured meat, the extracellular matrix will play an essential role in forming the texture and mouth-feel we associate with a juicy steak or a tender chicken breast.

In this journey, the cells will also need a source of energy to fuel their growth and multiplication. This is where mitochondria, the powerhouse of the cell, come in. No, really, they are often referred to as powerhouses because they generate most of the chemical energy needed to power the biochemical reactions of the cell. If this cellular city had a power plant, the mitochondria would be it. They'll be the ones keeping the lights on and the engines running in our meat cells.

Another key player in our cellular city is the membrane, a protective barrier that separates the inside of the cell from the outside world. But this isn't just a simple wall; it's a dynamic boundary that controls what enters and exits the cell, like an ultra-selective nightclub bouncer. It will keep our growing meat cells safe and sound while they're maturing.

And finally, let's not forget the lysosomes, often described as the recycling center of the cell. They break down waste materials and cellular debris, turning yesterday's trash into today's treasure. In the context of our cultured meat, they'll help keep the cells clean and healthy as they grow and multiply.

And there you have it, a crash course in Cellular Biology 101. As you see, it's not just small—it's a whole microscopic world down there! By understanding these fundamentals, we can begin to grasp how we might be able to coax a humble cell into becoming a juicy piece of meat, no animal farming required. From here, the Petri dish is your oyster—or rather, your oyster mushroom, your scallop, or your juicy beef patty. The possibilities are, dare I say, deliciously endless.

From Biopsy to Burger: The Process

And now, we find ourselves poised at the edge of the Petri dish, peering into the enchanting ballet of cellular life. The cellular city we've just acquainted ourselves with is about to embark on a journey—a journey from a mere biopsy to a bona fide, sizzle-ready burger. This process is a combination of science, technology, a bit of culinary arts, and a good dash of patience. Buckle up—it's time to talk process.

Our journey begins with a small, non-invasive biopsy from a living animal. This could be a cow, chicken, pig, or even a fish. It's a bit like taking a pinch of dough from a massive, living bread loaf, except the pinch is tiny, and the bread loaf has hooves (or fins, or feathers). The animals aren't harmed in the process, and one could even argue they enjoy the extra attention.

Once we have our biopsy, it's time to isolate those aforementioned marvels of adaptability—the stem cells. Picture them as our raw, unsculpted marble, just waiting for the touch of a microscopic Michelangelo. This step is a bit like panning for gold, but instead of gold nuggets, we're after cellular dynamos brimming with potential.

These stem cells are then placed in a nutrient-rich culture medium, their home for the next few weeks. Think of it as the stem cells' personal all-you-can-eat buffet, filled with a smorgasbord of sugars, amino acids, minerals, and vitamins. But this isn't just a feeding frenzy—it's the nurturing ground where the cells multiply and grow.

After a while, it's time for the stem cells to show their true colors (well, not literally, as that would be quite a psychedelic burger). They start differentiating into muscle cells. Guided by the carefully tailored conditions in the culture medium, our stem cells take on their meaty destiny. It's the equivalent of our cellular chameleons deciding to all dress up as Arnold Schwarzenegger for Halloween.

As the cells multiply and differentiate, they naturally begin to form small strands of muscle tissue. These are called myotubes, the building blocks of meat. This is where the magic happens—where cells become tissue, and tissue begins to resemble the familiar fibrous structure of the meat we know and love.

But meat isn't just muscle. To truly mimic the texture and flavor of conventional meat, we need fat. And so, another batch of stem cells is guided to become fat cells. It's a bit like casting the perfect supporting actor to bring out the best in our leading muscle cells. Together, they create a harmony of texture and flavor that is the trademark of a truly delicious piece of meat.

Now, our aspiring cultured meat is starting to come together, but it's still lacking structure. In traditional meat, this is provided by the animal's skeleton, but in our lab setting, we need to improvise. Enter the scaffolding—edible structures that help our meat cells align and merge into a 3D network. This might be made of a variety of materials, from plant-based substances to edible polymers. It's like the difference between a flat sheet of paper and a pop-up book, and boy, is it a page-turner.

On this scaffold, the myotubes and fat cells are combined and allowed to mature. As they grow, they naturally fuse into a structure that is, at a microscopic level, strikingly similar to the meat you'd find in a supermarket. But the petri-dish-to-pan transformation isn't quite finished.

This immature meat is then put through a process called 'bioreactor cultivation'. Now, despite sounding like a sci-fi term, a bioreactor is basically a fancy name for an environment-controlled chamber that provides the ideal conditions for our meat to mature. Think of it as a high-tech nursery for our budding baby burger.

And voila, what started as a tiny biopsy is now a cultured meat product, ready to be harvested, cooked, and enjoyed. It's a fascinating process, one that takes the miracle of life and directs it towards a new, sustainable, and ethical form of food production. The journey from biopsy to burger may be complex, but then, so is the taste of a perfectly grilled patty on a summer's day. And that, I assure you, is no mere meaty metaphor.

Innovations in Biotechnology

Welcome to the world of biotechnology innovation, a realm where scientists don lab coats instead of capes, and pipettes take the place of magic wands. Here, the impossible is not merely possible, it's Tuesday's agenda. With our knowledge of cells fully stocked and our understanding of the lab-grown meat process on point, we can now delve into the cutting-edge technologies that are shaping the future of our dinner plates.

First off, let's talk about 3D bioprinting. It's a term that sounds more at home in a Star Trek episode than a food science text, but the future, my friends, is a curious place. With this technology, cells, biomaterials, and biological molecules are used as "ink" to print, layer by layer, a 3D structure. The goal? To create complex tissue-like structures that mimic the real deal. Picture a meaty version of a 3D printed Eiffel Tower, if you will. This technology is enabling the production of lab-grown meat with a complexity and texture that was previously unattainable.

Next, let's visit the land of serum-free culture media. Earlier, we discussed the all-you-can-eat buffet that nurtures our growing cells. Traditionally, this buffet included fetal bovine serum—a product not without ethical and scalability concerns. But thanks to recent innovations, alternatives are being developed that are plant-based or synthetically produced. Imagine the switch from a controversially sourced buffet to a vegan-friendly feast, and you've got the idea.

In our journey through the hallways of innovation, we stumble upon a game changer: automation. Lab work can be tedious (just ask any graduate student), but automation is stepping in to take over repetitive tasks. Robots don't mind pipetting for hours or monitoring cell growth—they don't get bored or tired. This automation leads to greater accuracy, increased scalability, and the potential for cost reductions. It's like having a factory line for miniature burgers, if the factory line had a degree in cellular biology.

But innovation isn't only about shiny new technologies; sometimes it's about new applications for existing ones. Case in point: induced pluripotent stem cells (iPSCs). Originally developed for medical research, these are mature cells that have been 'reprogrammed' to behave like stem cells. In the context of lab-grown meat, they offer a potential renewable cell source, removing the need for repeated biopsies. It's like being able to hit the reset button on a cell's career path and telling it, "You know what? Why don't you try being a steak this time?"

Speaking of making the old new again, let's not forget the advances in bioreactor design. These are the high-tech nurseries where our lab-grown meat matures. New designs are being explored to increase efficiency and scalability, from wave-like motions to mimic the conditions inside an animal's body, to modular setups for easy expansion. It's like the difference between a static, old-school classroom and a dynamic, interactive learning space—both can educate, but one does it a bit more effectively.

Next on our innovation tour is gene editing. Tools like CRISPR are enabling scientists to make precise edits in the genome of cells. This could be used to enhance desirable traits, like growth rate or fat content, in our lab-grown meat. But don't worry, we're not talking about creating a Frankenstein's monster steak, just tweaking what nature gave us, like giving the cells a slight upgrade.

And then, there's the exciting world of cellular agriculture beyond meat. Scientists are looking at lab-grown versions of everything from milk and eggs to leather and silk. Each brings its own challenges and requires unique technological solutions. Imagine, a chicken-less egg or a cow-less leather jacket. Who said we couldn't have our egg and eat it too?

Last but not least, there's the realm of data analysis and machine learning. These tools allow us to sift through vast amounts of data from experiments and find patterns, predict outcomes, and optimize processes. It's like having a super-smart assistant who can see things you might miss, and learn from them. The age of data-driven dinner is upon us.

From 3D bioprinting to machine learning, the field of biotechnology is brimming with innovations that are pushing the boundaries of what's possible in lab-grown meat. Each discovery, each breakthrough, brings us a step closer to a world where our meat is not just tasty, but also sustainable and ethical. So, here's to the innovators, the dreamers, the bio-tinkerers—they're not just making a meal, they're crafting the future. And it looks delicious.

Chapter 3: Sustainability and Ethics

As we wade into the marshlands of sustainability and ethics, let's not forget our rubber boots, because things can get a bit muddy here. We're navigating a landscape where the carbon hoofprint of traditional meat production stomps heavily, leaving its indelible mark on our planet. On the other hand, we have the fresh sprout of cultured meat, promising a greener pasture. Yet, this isn't just a tale of two meats. There are ethical furrows to plow too, considering the promise of a future where our Sunday roast need not cast a shadow on animal welfare. Nevertheless, the race is not always to the swift, nor the battle to the strong, but that's the way to bet. Hold onto your hats, or rather, your toques blanches, as we dig into the sizzling intersection of lab-grown meat, sustainability, and ethics.

The Environmental Impact of Traditional Meat Production

Before we put the cart before the horse—or rather the beef burger before the cow—let's unpack the environmental carry-on of traditional meat production. Not exactly lightweight, I must say.

So, let's begin with the space required to rear our dinner. The grazing land needed for livestock isn't exactly petite. Imagine, if you will, dedicating an area the size of Africa to the task. Not just any old bit of Africa, mind you, but the whole Sahara Desert, the Serengeti, and the Nile Delta thrown in for good measure. And that's not even accounting for the vast acreages needed to grow the crops that feed the animals. We're playing a vast, high-stakes game of agricultural Tetris here.

Then there's the water story. Oh, the water. Picture filling an Olympic-sized swimming pool. Now imagine filling 15,500 of them. That's about how much water it takes to produce a ton of beef. It's enough to make you thirsty just thinking about it. But wait, that's not all. For each kilogram of beef, we need 15,000 liters of water. If you fancy a visual, that's like keeping your shower running continuously for 98 days straight. Talk about a long bath!

Let's not forget about greenhouse gases. Livestock farming contributes about 14.5% of all human-caused greenhouse gas emissions, according to our friends at the UN's Food and Agriculture Organization. That's more than all the cars, trains, planes, and ships combined. Indeed, cows could very well be the new Hummers.

Speaking of cows, they're a rather gassy bunch. No need to blush, we're all friends here. The simple act of digestion in ruminant animals results in the production of methane, a potent greenhouse gas. One cow alone produces the methane equivalent of a car burning through 235 gallons of gasoline in a year.

Our story wouldn't be complete without touching on the grand issue of waste. Not to tread in unsavory territory, but it's a fact that animals produce manure. Lots of it. This waste can contribute to water and air pollution, especially when not managed properly. Consider that a single dairy cow produces about 120 pounds of wet manure per day, equivalent to 20–40 people. An unwelcome party invitation if there ever was one.

Now, let's add a pinch of deforestation to the mix. Forests are often cleared to make way for pastures or to grow feed for livestock. It's like trading in your lungs for a hamburger—nobody's idea of a fair trade. And as if losing our forests wasn't bad enough, this process also releases vast amounts of carbon dioxide into the atmosphere, which is not what you'd call "helpful" in the fight against climate change.

Yet, here's the beef: all this isn't about pointing fingers or demonizing farmers. Many are just trying to make a living, often under tough conditions. But rather, it's to underscore the gravity of the situation and to highlight the need for better, more sustainable solutions.

At this point, you might be wondering if there's a 'but' coming, and indeed there is. This hefty environmental price tag doesn't have to be our only option. We've been savvy enough to invent selfie sticks, cloud storage, and instant coffee; surely we can muster something up for this meaty predicament.

After all, as we delve deeper into the subsequent chapters, you'll find that there's more than one way to sear a steak. But for now, let's ruminate on the environmental toll of our traditional meat production. It's a meal for thought that might just leave us hungry for change.

Cultured Meat: A Greener Alternative?

Having navigated the turbulent seas of traditional meat production's environmental implications, let's cast an eye towards calmer waters. Let's consider cultured meat. Lab-grown, cell-based, vat-harvested—call it what you will. The question is, could this meat sans the mooing and oinking be a greener alternative?

By design, cultured meat eschews the vast pastures, swarms of water guzzlers, and rather flatulent ruminants we've associated with traditional meat production. In this alternate universe, we're talking about bioreactors, nutrient-rich solutions, and cell cultures. It's like swapping the wild west for a high-tech lab worthy of a sci-fi flick.

First on our agenda: space. Lab-grown meat simply doesn't need the sprawling expanses of land currently dedicated to livestock. We're going from macro to micro, swapping hectares for petri dishes. It's like moving from a mansion to a studio apartment—cozier, yes, but it certainly has its perks.

Next up: water usage. When it comes to quenching the thirst of lab-grown meat production, we're talking more garden hose than Niagara Falls. An analysis by the think tank RethinkX suggests that cell-based meat uses 82-96% less water than conventional meat. So, you could say that cultured meat is more of a light sipper than a gulp-it-down kind of diner.

Now, about that pesky issue of greenhouse gases. While it's true that the production of cultured meat still generates some emissions, mostly from the energy used in the process, it sidesteps the methane problem entirely. No digestive systems involved, you see. The result? Some studies suggest a potential reduction in greenhouse gas emissions by up to nearly 90% compared to conventional beef. Not too shabby, I dare say.

Let's circle back to our less glamorous, but no less significant, topic: waste. In the world of cultured meat, there's no need to tip-toe around manure lagoons or deal with the associated methane emissions. In fact, the waste produced is primarily spent growth media—nutrient-rich liquids that could potentially be recycled. Not exactly dinner table conversation, but a bonus nonetheless.

We've saved the rainforests from our menu as well. With cultured meat, there's no need to fell forests to create pasture or cropland for animal feed. It's an approach that could spare an area twice the size of Australia from deforestation, according to some estimates. The trees breathe a sigh of relief.

But let's not get carried away on a unicorn here; lab-grown meat isn't without its environmental considerations. For starters, the energy required for its production, depending on the source, could still contribute to greenhouse gas emissions. There are ongoing debates and studies about the total energy input and how green it can actually get.

Then, there's the production of the growth media—a soup of sugars, amino acids, and minerals necessary for the cells to grow. This, too, requires resources. Yet, developments in biotechnology suggest that we may soon be able to create these inputs more sustainably, possibly even using carbon dioxide and renewable energy. Exciting stuff, isn't it?

Finally, it's important to remember that we're still in the early days of cultured meat. Many of the processes are not yet running at full efficiency, and much more research is needed. The good news is that improvements are being made, driven by the same innovative spirit that turned "cellular agriculture" from a geeky term into dinner table vernacular.

In the grand scheme of things, the simple truth is this: we need to feed a growing population without turning our planet into a well-done steak. Cultured meat, while not a silver bullet, seems to be a promising piece of the puzzle. A greener alternative? It's certainly shaping up that way. But, as they say, the proof of the petri-dish pudding is in the eating. Let's see where science leads us next.

Ethical Considerations and Animal Welfare

With a keen eye turned towards the future of our dining tables, it's time to tackle another piece of the cultivated cuisine puzzle: ethics. Yes, ethics! Those pesky questions of right and wrong that have haunted philosophers, theologians, and anyone who's ever sneaked the last cookie from the jar.

First, let's address the cow—or chicken, or pig—in the room. Traditional livestock farming, as we've seen, isn't exactly a garden party for the animals involved. Industrial agriculture has led to conditions in which animals are often raised in confined spaces, with limited access to the outdoors or opportunities for natural behavior. Meanwhile, cultured meat, which begins life as a small cell sample, involves no such confinement.

Which brings us to a question that might have been a topic of debate among cells, had they the capacity for such discourse: Is it unethical to harvest cells from animals? In the case of cultured meat, this typically involves a small, non-lethal biopsy. The animal, you might be relieved to know, gets to walk—or waddle, or strut—away afterward.

This "harvest" might be compared to taking a blood sample from a human for medical reasons, although some might argue that animals can't give consent. However, remember that these cells, once harvested, have the potential to produce far more meat than the animal could provide in its lifetime.

Now, onto the next hot potato: Are we playing Mother Nature here? Some people are uneasy about the idea of "unnatural" meat. They feel that messing with the building blocks of life crosses an ethical line. Yet, we've been selectively breeding livestock for centuries to enhance desired traits—a practice that one could argue is already meddling with nature.

Moreover, the practice of creating life in petri dishes isn't unique to cultured meat. We've been using similar techniques to produce insulin for diabetes patients and yeast for bread and beer. If you're going to be upset at meat grown in a lab, you might have to give up your daily bread and weekend pint, too. And let's be honest, nobody wants that.

Speaking of life in petri dishes, some might wonder if the cells used to make cultured meat are sentient. Well, despite what some sci-fi movies might have you believe, cells in a bioreactor aren't capable of suffering or experiencing life as we know it. They're too busy multiplying and differentiating, unaware of their destiny as a future steak or sausage.

However, let's not pretend that cultured meat will solve all animal welfare issues. We can't ignore the fact that traditional animal farming also serves purposes beyond meat production. Animals provide milk, eggs, wool, and work, and play vital roles in some ecosystems. Erasing this role could lead to unintended consequences.

Similarly, on the ethical front, the introduction of lab-grown meat brings up questions of food justice and sovereignty. Who will control the production of cultured meat? Will it be accessible and affordable for all, or will it become another way technology increases disparities in our food systems?

And finally, how does lab-grown meat impact our relationship with food and the natural world? Eating is a cultural, often emotional act tied to tradition, identity, and ethics. Cultured meat pushes us to re-evaluate these ties, to question what is natural, desirable, and morally acceptable when it comes to our diet.

What we can say is this: As with any technological advancement, the cultivation of meat brings with it a whole new menu of ethical considerations. And while we may not have all the answers yet, it's crucial that these questions are part of the conversation as we move forward. After all, progress tastes better when it's served with a side of thoughtful deliberation.

Chapter 4:
Health and Nutrition

Shimmy up your reading glasses and check your daily fiber intake, because we're about to dive deep into the nutritional landscape of lab-grown meat. We'll scrutinize the health implications from every angle, comparing this 21st-century marvel with its pasture-raised counterpart, and uncover the potential nutritional benefits that might have us swapping our backyard grills for bioreactors. While we'll also address any health concerns that could be hiding in the petri dish, rest assured—no microscopes or PhDs in biochemistry are required for this journey. Buckle up, and hold onto your B12 vitamins, we're delving into the world of health and nutrition, cultivated cuisine style.

Comparing Lab-Grown Meat to Conventional Meat

In this corner, weighing in at a whopping 300 billion pounds annually worldwide, we have the reigning heavyweight, conventional meat. And in the opposite corner, a newcomer strutting onto the scene with swagger, ambition, and a shiny laboratory pedigree, ladies and gentlemen, lab-grown meat!

Before we dive into this face-off, let's remember this isn't a question of good versus evil. No, no, it's more nuanced than that. It's more like comparing a 1960s vintage car with a modern electric vehicle. Both have their appeal, yet they are products of very different circumstances and technologies.

But first, let's consider the training regimen of our seasoned contender. Traditional meat production, a process as old as civilization itself, begins with an animal being raised, typically in a farm setting, then being sent to slaughter. The meat is then processed and distributed to our supermarkets, restaurants, and butcher shops.

This method, while tried and true, has room for variation. Our farm-raised protagonist can be a free-roaming grass-fed bovine, basking in the sunlight and chewing on the cud, or a grain-fed critter spending its days in a more confined setting. And believe me, there's a world of difference between the two when it comes to the end product.

Now, let's meet our bright-eyed challenger. Lab-grown meat, also known as cultured or cell-based meat, takes a few cells from an animal—let's imagine a rather surprised cow—and nurtures these cells in a nutrient-rich broth until they grow and multiply. The result is real, edible meat, no slaughter required.

Aesthetically speaking, our two contenders might look remarkably similar. Give a steak from each method a sizzle on the grill, serve it up with some potatoes and a side salad, and you'd be hard-pressed to tell which is which. Of course, we're only just beginning to explore the gastronomical potential of lab-grown meat, so stay tuned for future developments.

From a health perspective, both types of meat can deliver the nutrients your body craves – protein, iron, and essential vitamins. However, lab-grown meat may just take the edge, with the potential to be tailored for nutritional content. Picture this: a lab-grown steak fortified with omega-3 fatty acids. Sounds fishy? Perhaps, but the potential is certainly exciting.

In terms of environmental impact – a topic we've scrutinized in the previous chapter – lab-grown meat holds promise. With its potential to use fewer resources and emit fewer greenhouse gases, this innovative challenger could very well be the eco-friendly warrior of the future food arena.

Now, let's address the beefy elephant in the room, cost. As of now, conventional meat still has the upper hand here. However, like any fresh-faced contender, lab-grown meat is still finding its feet. As technology evolves and scales up, prices are expected to drop. So, let's not count our challenger out just yet.

Finally, there's the taste test, a category that continues to be hotly contested. Early reviews of lab-grown meat have been positive, but let's face it, your Grandmother's Sunday roast sets a high bar to clear.

In conclusion, we can see that lab-grown and conventional meat each bring their unique flavor to the table. While conventional meat has centuries of tradition and culture on its side, lab-grown meat represents a new paradigm with huge potential. As with all good face-offs, only time will tell which contender will dominate our dinner plates. For now, however, let's enjoy the ride and remember to pass the gravy.

Potential Nutritional Benefits

When we think about the potential nutritional benefits of lab-grown meat, it's a bit like imagining we've been given the keys to Willy Wonka's chocolate factory. Except in this case, instead of a river of chocolate, we have vats of nutrient broth, and rather than candy that makes you float, we're envisioning steak with personalized nutritional profiles.

The promise of lab-grown meat isn't just about providing a sustainable and ethical source of protein. It's also about the exciting possibility of fine-tuning the nutritional composition of our meat. Imagine for a moment, a world where your dinner not only tastes great but is also designed to meet your specific dietary needs. No, we're not talking about a magic pill or a meal in a shake. We're talking about lab-grown meat.

So, what's the nutritional scoop on our cultured contender? Well, the base-line here is that lab-grown meat, like its traditional counterpart, can provide a rich source of protein, as well as essential nutrients like iron and B vitamins. However, the nutrients aren't the end of the story; the fat content is where things really start to get interesting.

Traditional meat contains both saturated and unsaturated fats. Now, we've all heard the chatter about 'bad fats' and 'good fats', right? Generally speaking, eating too much saturated fat can increase levels of 'bad' cholesterol, while unsaturated fats can help increase 'good' cholesterol. In a traditional steak, the balance of these fats isn't something we can control. It's like trying to change the weather – intriguing to contemplate, but beyond our reach.

But with lab-grown meat, the fat content isn't dictated by nature. It's determined by the scientists who craft the nutrient broth and guide the development of the meat cells. It's not exactly like being a weather god, but it's the nutritional equivalent.

Scientists could potentially alter the fat content of the meat, reducing the amount of saturated fat and increasing the unsaturated fats. Imagine that! A steak that's low in 'bad' fat and high in 'good' fat. It's like having your cake and eating it too, but in this case, it's a steak.

Moreover, there is also the tantalizing potential to enrich lab-grown meat with specific nutrients. Imagine, if you will, a juicy, sizzling steak rich in heart-healthy omega-3 fatty acids. That's right, the same omega-3s that are commonly found in fatty fish and flax seeds might one day be a highlight in your beef steak or chicken breast.

And the possibilities don't stop there. Think about customizing meat to cater to specific dietary needs. A burger enriched with calcium for those who can't consume dairy, a steak boosted with vitamin B12 for vegetarians who occasionally indulge, or high-iron mince for the anaemia-prone.

Now, all this being said, let's not get carried away. These developments aren't just around the corner; they're a little further down the road. Scientists are still learning how to effectively and economically grow meat in the lab. Tweaking the nutritional content is a challenge for tomorrow.

And of course, lab-grown meat will never be a magical silver bullet for all our dietary needs. Eating a variety of foods, from fruits and vegetables to whole grains and lean proteins, is key to a balanced diet. But it's nice to imagine that, one day, our steak could be more than just a tasty piece of meat.

In conclusion, the potential nutritional benefits of lab-grown meat offer a tantalizing glimpse into the future of food. As the science advances, we might see a whole new culinary landscape unfold, where our dietary needs can be met in creative, sustainable, and delicious ways. And isn't that a future worth salivating over?

Addressing Health Concerns

In every great breakthrough, there's a chorus of concern. When the telephone was invented, people worried about losing the art of letter writing. When television emerged, people feared for radio's survival. And so, it's no surprise that as we prepare to welcome lab-grown meat into our homes and onto our plates, there are a number of health concerns brewing. It's like bringing a new pet home. Will it chew up the furniture or will it become our most faithful companion? Let's address some of those concerns now, with the verve of a dedicated veterinarian.

One of the primary concerns revolves around safety. The idea of eating something grown in a lab can be unnerving. It sounds too science-fiction, like teleporting to work or cleaning your house with a wave of a wand. But remember, we're not in the business of creating frankenfoods here. The process of producing lab-grown meat is monitored and controlled every step of the way. It's like the ultimate sous-chef, ensuring the meat is grown under the most sanitary conditions.

Of course, monitoring and control don't automatically equate to safety. As we know, even the most diligent of sous-chefs can have a bad day. So, rigorous testing is vital. But fear not, for our men and women in lab coats are more thorough than a detective with OCD. Extensive safety testing is undertaken before any lab-grown meat makes its way onto your plate. It's like an obstacle course, and only the safest meats make it to the finish line.

Another worry that crops up is about the use of antibiotics in the process. In traditional animal farming, the use of antibiotics is a significant issue. It's like a high school party that gets out of control, with antibiotics playing the role of rowdy teenagers. Overuse can lead to antibiotic resistance, a serious global health concern.

However, lab-grown meat presents an interesting twist in this tale. Theoretically, because of the controlled and sanitary environment in which the meat is grown, there's no need for antibiotics. It's a bit like growing plants in a hermetically sealed greenhouse where pests can't get in. You wouldn't need pesticides, right? In theory, lab-grown meat could be free of the antibiotic concerns that plague traditional meat production. But as we all know, theory and practice are like distant cousins. They're related, but they don't always get along.

So, while lab-grown meat has the potential to eliminate the need for antibiotics, rigorous standards and thorough testing will be required to ensure this remains the case. The process is like having a meticulous housekeeper who constantly checks for dust. We need to be sure no sneaky antibiotic issues find their way in.

Moving on to the question of allergenicity. As with any new food, there's the potential for unexpected allergic reactions. It's like introducing a new species into an ecosystem - we can't predict all the interactions. In the case of lab-grown meat, it's vital to monitor for new allergens, but remember, the cells used to grow the meat come from animals that we already consume. It's not like we're cooking up a steak from a newly discovered alien species.

Lastly, there's the question of long-term health effects. Since lab-grown meat is a new entrant to the culinary scene, we simply don't have long-term data on it yet. It's like trying to predict the career of a newly drafted quarterback. Sure, the potential is there, but we'll just have to wait and see how it plays out.

In conclusion, as with any food, lab-grown meat must pass rigorous safety and health tests before it makes its debut on our plates. While potential health concerns are valid and must be addressed, the science, the potential, and the promise of lab-grown meat present a compelling argument. Just as we embraced the telephone and the television, we may one day find ourselves reminiscing about our initial fears of lab-grown meat while we enjoy a lab-crafted steak at our favorite restaurant.

Chapter 5:
Overcoming Challenges

In the grand scheme of culinary revolution, lab-grown meat finds itself in the role of an unassuming hero, like a mild-mannered accountant suddenly called to save the world. The challenges it faces are nothing short of daunting, ranging from technological hurdles fit for a sci-fi movie, economic viability puzzles that could make an economist tear their hair out, to legal and regulatory landscapes more intricate than a game of chess with a supercomputer. But, as with any hero, it's not about the size of the challenge, but the determination to overcome it. In the forthcoming sections, we'll dive into these tribulations, wearing the mantle of intellectual curiosity, humor at the ready, and armed with the power of science. After all, the steaks have never been higher.

Technological Barriers and Scientific Hurdles

Let's kick this off by acknowledging the elephant in the room, or more fittingly, the cow in the lab. Creating meat without raising and slaughtering an animal is a bit like creating a sculpture without touching a chisel to marble - it sounds like a plot twist from a science fiction novel. And yet, here we are, the scientific community, standing at the precipice of making the seemingly impossible a reality.

One of the primary barriers in this scientific narrative is the culture medium, the nutritional soup that coax cells to grow into meat. For cultured meat to be a sustainable alternative to traditional meat, the culture medium must be affordable, animal-free, and scalable. At present, this concoction often includes fetal bovine serum, a product derived from the blood of unborn calves - not exactly the cruelty-free narrative we're striving for, and the cost can make even the most deep-pocketed investors wince.

A second barrier is the scalability of the production process. Growing a few cells in a petri dish in a lab is one thing, but scaling that to produce enough meat to satisfy the appetites of a planet's worth of burger lovers is another matter entirely. It's like trying to fill a swimming pool with a teaspoon. Sure, it's technically possible, but we'd all prefer a more efficient method.

Then, there's the challenge of muscle tissue. Meat, as we currently know it, gets its texture from a complex structure of muscle and fat. Muscle tissue, in particular, needs to be exercised to develop correctly. If you think your gym routine is tough, spare a thought for these lab-grown cells. Without the equivalent of tiny cell-sized treadmills, the challenge lies in figuring out how to encourage the cells to form realistic muscle tissue.

Right, onto the fat. A juicy steak owes much of its flavor to its marbling, the thin streaks of fat woven into the muscle. Achieving the right balance of muscle to fat cells is another hurdle that needs to be crossed. Without it, we might end up with a product closer to a rubber eraser than a filet mignon.

You might now be asking, "Can't they just grow an entire steak as is?" and I applaud your ambitious thinking. But that brings us to the issue of vascularization - in layman's terms, the network of blood vessels that delivers nutrients to cells. So far, we can grow thin slices of meat, but anything thicker requires a way to distribute nutrients evenly throughout the tissue. Creating such a network is an intricate ballet of biology and engineering.

Closely linked to the previous point is the challenge of bioreactors - the apparatus in which our cells will be grown. Imagine trying to find a home for millions of fastidious would-be steaks. These bioreactors need to offer conditions that mimic a cow's body: right temperature, correct pH, and perfect nutrient supply, among other things. Finding the Goldilocks zone of perfection is easier said than done.

Lastly, we can't overlook the issue of time. Current processes for producing cultured meat are time-consuming, and time, as they say, is money. It can take weeks to produce a small amount of lab-grown meat. By comparison, traditional farming might seem as quick as a microwaved meal, and for this technology to be commercially viable, we need to turn the tables on Father Time.

In the face of these challenges, it's easy to feel daunted. After all, we're not just re-inventing the wheel here, we're re-inventing the steak. But it's crucial to remember the profound potential benefits that this technology can bring. If successful, we stand to transform one of the most significant sectors of the global economy, and in doing so, address some of the most pressing environmental and ethical issues of our time.

So, let's not consider these barriers as stumbling blocks but as stepping stones. Because in science, as in life, it's the challenges we overcome that define our progress. And if these hurdles seem high, just remember, we're aiming for the moon, or in this case, the meat.

Economic Viability and Scaling Production

Economic viability is to a business what gravity is to an apple - it's the force that grounds it or, in a less fortunate scenario, leads to a substantial splat. Let's delve into the subject of the economic viability and scaling production of our lab-grown meat venture, shall we?

Presently, producing lab-grown meat involves an array of costly inputs, from the culture medium to the energy required to keep the bioreactors humming along. It's like throwing a dinner party where the guests are cultured cells and their appetites are voracious. The costs involved can make the price tag of cultured meat more akin to a luxury product than your everyday sirloin.

But this isn't the whole story. The beauty of technology is that it tends to get cheaper over time - we've seen this with everything from computers to electric cars. Imagine if lab-grown meat followed a similar trajectory. Today's filet mignon price might be tomorrow's ground beef. As production methods improve and economies of scale kick in, costs will likely decrease.

Economies of scale, by the way, is the principle that the cost per unit decreases as the quantity of production increases. It's like buying in bulk at the supermarket - the more you buy, the cheaper each item becomes. In the case of lab-grown meat, this could mean that as production ramps up, the price per pound of cultured meat may drop significantly.

But let's not get ahead of ourselves. Increasing production brings its own challenges. This isn't merely a case of turning up the dial on the bioreactor. Remember our earlier discussion on the complexity of the culture medium, muscle and fat structure, vascularization, and the Goldilocks condition of the bioreactors? Scaling up production must negotiate all these scientific hurdles while maintaining consistency and quality, a task easier said than done.

While we're in the territory of challenges, let's also consider the resource input. Water, electricity, physical space - these are all necessary for production. Sure, preliminary studies suggest cultured meat could use less water and land than conventional meat production, but it might need more energy. So, getting the balance right to ensure a more sustainable process while keeping costs down will take some fine-tuning.

There's also the matter of infrastructure. Transitioning from traditional farming to cultured meat production will require new facilities, equipment, and trained personnel. This transition might be comparable to shifting from typewriters to computers – a revolution of meaty proportions.

The role of government and policy shouldn't be overlooked either. Incentives, subsidies, and supportive legislation could help to accelerate the development and adoption of lab-grown meat, just as they've done for renewable energy. The flip side? Restrictive regulations or public pushback could slow things down. Navigating the policy landscape will be as crucial as the scientific one.

And let's not forget about the competition. Traditional meat isn't going to go down without a fight, and plant-based meat substitutes are also vying for a piece of the protein pie. Success in the market will not just depend on being a viable alternative, but on being a desirable one. This isn't a race for the swift, but for the scrumptious.

All these considerations may seem daunting - and let's be honest, they are. But remember, every great innovation in history has faced its share of challenges. From the invention of the wheel to the development of the internet, each leap forward has had to negotiate economic realities, scaling challenges, and fierce competition.

In the end, the question of economic viability and scaling production of lab-grown meat will be answered by a complex recipe of science, economics, policy, and, most importantly, our appetite for innovation. The path is steep, but the potential reward is profound. Let's see if we can cook up a storm.

Legal and Regulatory Landscapes

Navigating the legal and regulatory landscapes of lab-grown meat is a bit like trying to play a game of chess on a roller coaster - it's confusing, occasionally nausea-inducing, and the rules seem to change every time you blink. Let's unpack the complexities of this thrilling ride, shall we?

The lab-grown meat industry is a newborn colt in the field of food regulations, and as you can imagine, this field is filled with the nettles of intricate legislation and policy. These nettles have the potential to sting or, if handled correctly, to be transformed into a nourishing soup of progress.

One of the primary regulatory concerns is safety. After all, we're talking about a product that people are going to eat. To ensure consumer safety, lab-grown meat must go through rigorous testing and approval processes before it can grace our supermarket shelves or restaurant plates.

At the heart of this safety dance are two key U.S. agencies: the Food and Drug Administration (FDA) and the United States Department of Agriculture (USDA). The FDA is responsible for ensuring the safety of biotechnology processes, while the USDA oversees meat and poultry inspection. Together, they're like the parental units of the food world, each with their unique but overlapping roles.

Now, consider that lab-grown meat is a kind of hybrid, falling under the purview of both agencies. This has resulted in a regulatory tug-of-war with each agency pulling to assert its authority. It's like a game of territorial twister, but with less laughter and more paperwork. The result is a unique regulatory framework where the FDA oversees cell collection and growth, and the USDA monitors production and labeling.

Labeling, by the way, is another hot potato. What exactly do we call this product? Lab-grown meat? Cultured meat? Cell-based meat? Each term carries its own connotations, and the choice of name could impact consumer perception and acceptance. The USDA has yet to finalize its labeling rules, leaving us in suspense like a cliffhanger at the end of a season finale.

However, this isn't just a U.S. story. Lab-grown meat is a global endeavor, and regulatory landscapes differ from country to country. The European Union, for instance, has its novel food regulation, while Singapore was the first country to approve the sale of lab-grown meat. How countries regulate lab-grown meat could significantly affect its global development and adoption. It's like a game of international regulatory poker, where the stakes are high, and the bluffing is part of the game.

But regulations aren't just about hurdles; they're also about opportunities. Positive regulations can support and accelerate the growth of the lab-grown meat industry, providing clear guidelines and fostering public trust. After all, a well-crafted rulebook is essential to any well-functioning game.

Despite the challenges, the wheels are turning. Policies are being drafted, rules are being established, and slowly but surely, the legal and regulatory landscapes are being shaped. It's a bit like watching the evolution of a complex organism, and we're in the primordial soup stage, with the promise of an exciting future creature.

So, while it may be an uphill trudge through dense legalese undergrowth and sometimes slippery policy slopes, the journey towards establishing a comprehensive legal and regulatory framework for lab-grown meat is underway. Like any good adventure, it's filled with challenges, unexpected turns, and the potential for significant rewards. Buckle up, because this regulatory roller coaster is just beginning its thrilling ride.

Chapter 6:
Consumer Perception and Market Adoption

L ike it or not, the advent of lab-grown meat has us on the precipice of a meat revolution, and we, the consumers, find ourselves at the frontline, armed with our utensils. This chapter is about how we, the brave infantry of the dinner table, perceive this nascent edible technology. We'll also delve into how lab-grown meat might successfully infiltrate our shopping lists and eventually, our plate. Will it strut down the aisles of our supermarkets or sneak in like a culinary ninja? It's a heady mix of science, psychology, marketing, and a good dash of culinary bravado. We're talking acceptance, we're talking disruption, we're talking the potential foodie revolution of the century. And believe me, this isn't your grandma's Tupperware party.

Public Opinion on Lab-Grown Meat

Public opinion on lab-grown meat is as diverse and varied as a late-night TV lineup - and just as subject to change based on the latest trends and performances. You have your comedy fans, the skeptics who wouldn't touch a Petri dish burger with a 10-foot grilling fork, and then the science fiction enthusiasts who've been waiting for this since they first read "Do Androids Dream of Electric Sheep?"

One of the most compelling and confounding aspects of lab-grown meat is that it blurs traditional ideological lines. The vegan who philosophically opposes the killing of animals for food might now be able to order a steak without any moral compromise - yet how many self-respecting vegans would be willing to take that leap? On the other hand, the meat-loving barbecue pitmaster might be drawn to the efficiency and environmental sustainability of lab-grown meat, but would they ever really feel comfortable replacing their beloved brisket with a lab-cultured equivalent?

Interestingly, survey says - well, various surveys say different things. For example, some polls indicate that a majority of Americans are willing to give lab-grown meat a try. Conversely, other surveys paint a less rosy picture, with large numbers of respondents expressing a "thanks, but no thanks" attitude to the prospect of lab-grown beef on their dinner plates. Evidently, it seems that public opinion, much like a fine soufflé, can be quite delicate and temperamental.

And what about taste? Flavor is, after all, a kingmaker in the world of food. Will the public warm up to the taste of lab-grown meat, or will they turn up their noses at this science-fueled feast? The jury is still out, but early taste tests indicate that the technology is indeed promising. Still, whether this promising start translates into widespread adoption is a question that only time, and a lot of taste buds, will tell.

Culturally, lab-grown meat faces another challenge: tradition. Culinary traditions, from the American barbecue to the Argentine asado, are about so much more than just food. They're about the experience, the history, and the cultural identity that comes with them. Can a slab of lab-grown beef really compete with the deep-seated love for these culinary rites of passage?

Then there's the yuck factor. The idea of eating something that was created in a lab is, to many, about as appetizing as eating a plastic fruit display. This is one of the biggest hurdles that lab-grown meat will have to overcome. It's not just a question of taste, but of perception. Lab-grown meat might be molecularly identical to conventional meat, but to many, it's the idea that counts. It's like telling someone they're eating a delicious apple, only to reveal after they've bitten into it that it's actually a genetically modified superfruit developed in a lab.

Yet, despite all these challenges, there's an unmistakable sense of optimism in the air. Early adopters and enthusiasts are already showing their support, and the environmental and ethical arguments are indeed compelling. It's not so much a question of "if" as "when". And when that time comes, it will be fascinating to see how the scales of public opinion shift.

To finish, here's a fun thought: Imagine going to a dinner party ten years from now. You're enjoying your meal when the host casually mentions, "By the way, the meat was grown in a lab." What's your reaction? A gasp? A shrug? A request for seconds? This is the crux of the matter. As we move forward, the question will become less about whether lab-grown meat can replicate the taste and texture of traditional meat, and more about whether society is ready to accept this new food frontier.

In the end, public opinion on lab-grown meat is a buffet of hopes, fears, curiosity, and skepticism. But as with any buffet, tastes can change. It's all part of the culinary adventure, and what an exciting adventure it promises to be. So, grab your fork and your open mind, and let's dig into the future.

Changing Food Culture

Welcome to the rollercoaster ride of changing food culture. Buckle up, because the twists, turns, and loop-de-loops are part of the journey when it comes to how society views, values, and consumes its food. In the blink of an eye, we've gone from being hunters and gatherers to microwaving 2-minute noodle dinners and ordering exotic fusion cuisine with a single tap on a smartphone. Now, lab-grown meat threatens to bring the biggest upset yet to the culinary table. It's like a surprise ingredient in a cooking show that has the contestants scrambling.

Consider the avocado, that humble fruit (yes, it's a fruit, and a berry no less) that somehow climbed the ranks to be the star of breakfast plates and Instagram feeds across the world. Its rise from obscurity was no accident, but the result of a concentrated and persistent effort to change how the public perceived it. A similar fate could await our lab-grown meats, a nouveau riche, if you will, in the culinary scene.

Lab-grown meat represents more than just a new type of food; it signifies a seismic shift in our relationship with what we eat. The central tenet of food culture has always been familiarity – the warmth of a home-cooked meal, the nostalgia evoked by family recipes handed down through generations. But the advent of lab-grown meat brings a jolt of the unfamiliar into this cozy scenario. A chicken without a cluck or a cow without a moo? It's enough to send Bessie spinning in her pasture.

However, it's worth noting that humanity has always been adaptable, especially when it comes to food. Sushi, for instance, went from being an oddity outside Japan to becoming a globally recognized and loved dish. Remember when raw fish was considered 'adventurous' eating? Now it's as mainstream as pizza on a Friday night. Could lab-grown meat eventually follow the same path from 'weird science' to 'weeknight staple'? Only time (and the brave pioneers of the dining scene) will tell.

Yet, the question remains: Can lab-grown meat truly find its place within our diverse culinary traditions? Will there be lab-grown ribs at the American BBQ, or lab-grown lamb on the Australian grill? It's a fascinating thought. I mean, if they can make a decent mock duck out of tofu, then a lab-grown burger has to be in the realm of possibility, right?

But it's not just about replicating existing dishes. Lab-grown meat also offers the potential for entirely new culinary creations. Imagine being able to tweak the fat content of your steak to get the perfect marbling or choosing the exact flavor profile of your chicken breast. The opportunities for customization are mind-boggling, like a meaty version of a pick-and-mix candy store.

Of course, all this change can be intimidating. And it's not just the consumer who is affected. The chefs, the artisans of our food culture, are also in the line of impact. Will they embrace this new ingredient or shun it? Will the culinary schools of the future have classes in 'Cultured Meat 101'? One can imagine a heated debate in the kitchen, with one chef insisting, "There's no replacing the real thing!" and another retorting, "But think of the possibilities!" It's like a culinary reality TV show waiting to happen.

Despite the challenges, it's clear that lab-grown meat is set to play a significant role in shaping our future food culture. Whether it's through necessity, curiosity, or a genuine desire to make more sustainable choices, there's a growing appetite for what lab-grown meat can bring to the table. It's a new twist in the ever-evolving saga of our culinary journey.

The finale? Well, that remains to be seen. Will lab-grown meat be the star of the show, or will it flounder like a poorly executed soufflé? Will it change the way we think about food, or will it be relegated to the annals of culinary history as an ambitious but ultimately unappetizing experiment? The only certainty is that it will be a fascinating watch. So, grab your popcorn (or should that be your lab-grown chicken nuggets?) and enjoy the show. This, dear food enthusiasts, is the future of our food culture.

Marketing and Branding Strategies

Marketing and branding strategies for lab-grown meat—now there's a topic that would give even the Mad Men a run for their money. In a market where your product is as unfamiliar as, say, a fish that enjoys mountain climbing, the right strategy could make all the difference between an enthusiastic welcome and a polite but firm "no, thank you."

Let's begin by discussing the elephant in the room—how to make lab-grown meat, an inherently scientific and, let's face it, slightly unappetizing concept, appealing to the average consumer. Traditional meat has the advantage of centuries of cultural tradition, of Sunday roasts and barbecue cook-offs. Lab-grown meat, in contrast, has the sterility of a petri dish. It's like inviting someone to a party but telling them it's in a laboratory.

So, the first hurdle in marketing lab-grown meat is crafting a compelling narrative, one that can imbue this new product with the warmth and familiarity associated with traditional meat. The food industry has done this before, turning breakfast cereal from chicken feed to a staple for humans. So why not transform lab-grown meat from a scientific oddity to a mainstream meal choice? It's not so much a pig in a poke as a steak in a flask.

This narrative is crucial because it can shape the very identity of the product. Take the name itself: lab-grown meat. It's rather scientific, don't you think? Terms like 'cultured meat' or 'cell-based meat' might make the product sound more sophisticated, but they still carry a whiff of the lab about them. Some companies are even exploring more appetizing names that emphasize the product's qualities. Who knows, we might end up with 'EcoBurger' or 'KindCutlet'. As long as it's not 'TestTubeT-bone', we should be fine.

Aside from the name, there's also the issue of how to represent the product visually. A juicy steak sizzling on a grill is a staple image in meat advertising. But how do you create an equally appealing image for lab-grown meat? One approach might be to emphasize the end result—show a cultured meat burger in all its glory, complete with all the trimmings. After all, we don't sell cars by showing people the factories they're built in, do we?

Another key element of the marketing strategy is targeting the right audience. Early adopters are likely to be younger, more environmentally conscious consumers—those who are comfortable with change and have a taste for the futuristic. It's the equivalent of selling an electric car; you wouldn't start by pitching it to the hardcore muscle car enthusiasts, now would you?

Partnerships also offer a significant opportunity. Collaborations with popular chefs and restaurants can lend credibility to lab-grown meat. Imagine a world-renowned chef offering a signature dish made with cultured meat. That's the kind of endorsement that could elevate lab-grown meat from a novelty to a delicacy. It's a bit like having a celebrity drive your new electric car—it adds an extra layer of desirability.

There's also a role for education in the marketing strategy. Lab-grown meat offers benefits in terms of sustainability, animal welfare, and potentially even health. Informing consumers about these benefits can help them see beyond the 'yuck' factor and understand why lab-grown meat might be a choice worth making. This isn't so much selling a product as selling a future—a future where we can enjoy meat without the associated guilt or environmental damage.

However, there's one last hurdle that marketing has to overcome: price. For all its benefits, lab-grown meat will only become a viable choice for most consumers if it's competitively priced. It's like trying to sell a diamond-encrusted smartphone—sure, it's fancy, but if it's too expensive, most people will stick with the plain version.

So, there you have it, the labyrinthine path that marketing and branding strategies for lab-grown meat will need to navigate. It's not an easy task, trying to sell a product that's so new, so different, and, to some, so outlandish. But then again, that's what makes it so darn interesting. After all, if marketing was easy, it wouldn't be nearly as much fun. And who knows, with the right approach, lab-grown meat might just become the next big thing. Stranger things have happened—just ask the avocado.

Chapter 7: The Global Impact

Now we're diving into the vast ocean of lab-grown meat's global impact. This chapter promises to be as exciting as watching the World Cup final while riding a roller coaster—backwards. Here, we'll chew on the potential effects on farmers and industry, probe into the international scene (spoiler alert: it's not just the mad scientists in the United States who are investing in this), and explore future scenarios ranging from meat shortages to a vegan world. Brace yourself for a whirlwind tour through the lab-grown meat's potential to redraw the map of global agriculture, transform our economic landscapes, and rewrite our cultural norms around food. Just think of it as the ultimate food fight —with the fate of our planet on the dinner plate.

The Potential Effect on Farmers and Industry

Enter the world of farmers and traditional meat industry—rich with history, full of quirks, and sometimes stinkier than a week-old, sun-baked haggis. When it comes to lab-grown meat, many from this realm might feel like they've woken up one morning to find that their horse and plow have been replaced by a teleporting tractor. Shocking, bewildering, and downright sci-fi. But let's lean into that sense of surprise, shall we?

If the meat industry were a movie, traditional farming would play the beleaguered protagonist. It's been the backbone of our civilization, the bread and... well, meat, of our society. Still, like any good cinematic hero, it's got its flaws. It's tough on the environment, as we've mentioned earlier, and sometimes not so great for our health. Enter cultured meat—the suave newcomer on the block, offering a shiny, eco-friendly alternative. But what does it mean for our protagonist?

Let's start with job displacement—an elephant in the room with a sizeable appetite. Yes, cultured meat may potentially reduce the need for traditional farming practices. However, there's a plot twist. It could also create a slew of new roles in biotechnology, cellular agriculture, and other science-fiction-sounding fields. Think of it as a career plot twist. Sure, it's disorienting, but it also opens up a whole new world of possibilities.

But hold onto your pitchforks, it's not all about job loss. There's also the matter of resource allocation. Traditional farming, especially livestock farming, requires vast tracts of land, substantial amounts of water, and, frankly, a lot of time. Cultured meat, on the other hand, is a bit of a minimalist. Less land, less water, less time. It's like the Marie Kondo of meat production, and it might just spark joy for our planet.

Now, let's talk about animal welfare, an issue that stirs the pot faster than a master chef. Lab-grown meat presents an opportunity to substantially reduce, if not eliminate, the need for factory farming, a practice often under fire for its ethical issues. For those who've had sleepless nights over the treatment of livestock, this could be a reason to finally hit the snooze button.

And then, there's the potential for economic disruption. The meat industry is, in technical terms, a big deal. It's a critical part of many economies worldwide, from the big guns like the United States and Brazil to smaller, agriculture-focused countries. An industry shift towards lab-grown meat could have ripple effects, much like dropping a pebble in a pond. Or, in this case, a meatball in a bowl of soup.

But it's not just about economics—it's also about health. One of the most important aspects of any food production system is its potential impact on public health. As we mentioned earlier, lab-grown meat could significantly improve food safety and reduce the risk of zoonotic diseases, which, spoiler alert, is a pretty big deal in the post-COVID-19 world.

However, as cultured meat saunters onto the scene, it's likely to encounter resistance. We humans are creatures of habit. We like our burgers grilled, our steaks seared, and our chicken fried. Acceptance of this new form of meat will take time, but hey, even the microwave was considered a witch's cauldron at first.

Lastly, don't forget the influence of policy and regulation on farmers and the industry. The landscape of lab-grown meat is much like a wild west of regulatory issues. Who oversees its production, how it's labeled, how it's marketed—all of these will have significant impacts. Stay tuned as we navigate this new frontier.

So, what's the conclusion of this little movie of ours? Well, the potential effect of lab-grown meat on farmers and the meat industry is complex, multi-layered, and frankly, still a bit of a mystery. But like any good story, it's the journey, not the destination, that matters. Let's see where this one takes us.

The International Scene: Who's Investing?

Ah, the international scene—the global cafeteria where every country is queuing up to serve its slice of the lab-grown meat pie. It's a bustling, crowded, somewhat messy affair, as you might expect when the future of food is on the table. Now, who's bringing the largest trays and the deepest pockets? Let's peer over the rim of our lab glasses and see.

For starters, let's make a pitstop in the United States. The land of the free, the home of the brave, and increasingly, the playground for the cultured meat industry. The U.S. has been a major hub for this culinary revolution, with multiple start-ups springing up like petri dish mushrooms. These ventures, with names more suitable for sci-fi novels than butcher shops, are rapidly attracting investors, with millions—no, billions—of dollars flowing into their coffers.

Moving eastward, we alight in Israel, a nation that is punching well above its weight in the lab-grown meat stakes. Despite its small size (it could comfortably fit inside Lake Michigan), Israel is a veritable titan in cultured meat research and development. With groundbreaking companies like Aleph Farms and SuperMeat calling it home, it's a proverbial Silicon Valley of steak without the moo.

Next, we traverse the breadth of Eurasia to visit the land of the rising sun. Japan has thrown its proverbial hat in the ring and it's a hi-tech one at that. Government initiatives and corporate ventures are fostering an environment that's as friendly to lab-grown meat as a sushi chef to a bluefin tuna. Japanese corporations are already creating the first lab-grown wagyu, a concept as intriguing as it is appetizing.

Now, let's head to Singapore, where regulatory green lights have paved the way for the world's first commercial sale of lab-grown chicken nuggets. That's right, in this city-state, one can sit down in a restaurant and order a plate of lab-grown poultry, something that would have seemed as fantastical as riding a unicorn just a few years ago.

Not to be outdone, China is also entering the fray. With a population larger than the last ten Super Bowl crowds combined, China's interest in lab-grown meat could represent a significant shift in the global food industry. Some Chinese tech giants are already investing heavily in cultured meat, showing that this isn't just a passing fad, it's the moo-less moo-shu of the future.

Meanwhile, in Europe, a combination of innovation, regulation, and a growing concern for sustainability have put lab-grown meat firmly on the menu. In the Netherlands, Mosa Meat, whose scientific co-founder brought us the first lab-grown burger, continues to pioneer this exciting new field. It's the Van Gogh of vat-grown meat, if you will.

Australia, ever the competitor, is not far behind. Recent government funding and initiatives toward cell-cultured meat indicate that Australia wants a prime seat at the cultured meat barbie. Considering that Aussies eat their weight in meat every year (slight exaggeration, but you get the drift), it's a promising development.

But it's not all rosy. Some countries, particularly those with economies heavily reliant on livestock farming, are viewing this new tech with a suspicious eye. It's like someone just showed up at the barbecue with a kale salad—it's green, it's healthy, but it's not what they came for. There will be challenges and opposition, much like trying to flip a burger with a pair of chopsticks.

The investments pouring in globally aren't just about making a quick buck (or yen, or euro, or yuan). It's about the recognition that our current way of producing meat is not sustainable, not scalable, and in some ways, not ethical. Cultured meat offers an alternative that's too intriguing, too promising to ignore.

So, while the international scene of lab-grown meat investment might feel like a hectic rush hour in Tokyo, it's a rush towards a future that could be healthier, kinder, and tastier. Now, isn't that worth investing in?

Future Scenarios: From Meat Shortages to Vegan World

Oh, the future! That magical, mysterious realm where anything can happen. From dystopian wastelands to utopian wonderlands, our collective vision of what lies ahead is as diverse as a box of assorted chocolates. But enough metaphors, let's venture into the potential scenarios that may await us in the land of lab-grown meat.

Imagine, if you will, a world gripped by a meat shortage. Not a great starting point, I know, but bear with me. Traditional meat production is under strain, as climate change, water scarcity, and land degradation conspire to restrict output. Suddenly, your favorite Sunday roast is priced like a truffle-infused caviar. Cue cultured meat, swooping in like a lab-coated superhero. With the ability to produce meat without the constraints of traditional farming, it could keep our plates and palates satisfied, turning a potential crisis into just another dinner conversation.

Or consider this: a world where lab-grown meat is the norm, and the word "slaughterhouse" is as archaic as "dial-up internet." This is a realm where cultured meat has been so successful, so universally adopted, that traditional farming has been relegated to a niche practice. Like vinyl records in an era of streaming music, farm-raised meat becomes a quirky relic for those who prefer the old-fashioned ways.

But let's not stop there. Suppose we journey further down the rabbit hole to a vegan world—no, not a planet where animals rule and humans are kept as pets, but a world where all our meat is lab-grown. In this scenario, we would effectively be "vegan," eating meat without any animal suffering. It's a world where the term "bloodless steak" is not a culinary criticism, but a boast of ethical advancement.

In a somewhat darker avenue of the future, lab-grown meat might not manage to oust traditional meat. Instead, it could become an option for those who can afford it, while everyone else continues to consume regular meat. It's a future that mirrors our current reality with organic food—a premium choice for those with deeper pockets.

Then again, what if lab-grown meat outcompetes traditional meat on price? Imagine cultured meat being cheaper, more efficient, and just as tasty. Traditional meat could become a luxury item, the stuff of gourmet restaurants and high-end supermarkets, while lab-grown meat feeds the many. It's like finding out that the champagne of the future is made from potatoes—surprising, yet oddly satisfying.

However, let's not overlook the possibility of a backlash. Change, even when promising, can be scary. A world resistant to lab-grown meat, driven by fear, misinformation, or a steadfast clinging to traditional ways, is a plausible future. Think of it as the culinary equivalent of people refusing to give up their horse-drawn carriages in the age of automobiles.

On a brighter note, let's visualize a scenario where lab-grown meat aids in combating global hunger. If the production process can be streamlined and made affordable enough, cultured meat could help feed populations that currently lack consistent access to protein. Picture a future where "world hunger" is an obsolete term, as antiquated as a typewriter.

And what about the animals, you ask? Imagine a world where the livestock population shrinks, where vast pastures are freed up for rewilding or crop cultivation. It's a future where the phrase "free range" isn't a selling point, but a given, where the cow truly jumps over the moon, not because it's fleeing a slaughterhouse, but simply because it feels like it.

In all seriousness, these scenarios are speculative, of course. The future rarely adheres to our neatly plotted graphs and predictions. It has a knack for throwing curveballs, for surprising us in ways we couldn't even imagine. After all, who foresaw a time when we would be discussing lab-grown meat as a legitimate food source?

So, whether we're facing a meat shortage, entering a vegan world, or something in between, one thing is certain: the future of food is exciting, and lab-grown meat is set to play a starring role. But remember, it's not the food on our plates that matters most, but the people around our tables. And hopefully, they'll all be open to trying a bite of the future.

Chapter 8:
A Peek into the Future of Food

As we stand on the brink of this brave new world, armed with our petri dishes and pipettes, we're about to explore the food frontier beyond meat. Yes, that's right. Lab-grown meat was merely the opening act, and boy, what a show it's been. But wait, there's more. Our menu is expanding, with cultured dairy, eggs, and seafood warming up backstage, ready for their culinary debut. It's a tantalizing tango of science and gastronomy, a future where chefs become bioengineers and the dining table turns into a showcase of innovation. And, as we venture into this new era, we'll not only reimagine what we eat but also how, when, and where we eat it. So, hold on to your forks, folks, the future of food promises to be a culinary rollercoaster ride. Now, who's hungry for a taste of tomorrow?

Beyond Meat: Lab-Grown Dairy, Eggs and Seafood

As we tumble headlong into the epoch of cellular agriculture, the prospect of expanding the horizons beyond lab-grown meat becomes less of a pipedream and more of a logical step forward. Yes, indeed, it's time to turn our attention to lab-grown dairy, eggs, and seafood, or as we like to call it, the great food frontier that looks suspiciously like your grandmother's Sunday brunch.

Let's get this moo-ving (couldn't resist a good cow pun, could we?) with lab-grown dairy. The science behind cultured dairy parallels that of our trusty lab-grown meat, where cells are cultured and nurtured to develop into milk proteins. The result? Everything from cruelty-free milk to the cheese that graces your favorite pizza could be produced without a cow ever having to endure a milking machine. There's a potential "utter" revolution on the horizon, and it's not too cheesy to say we're excited about it.

Next up, let's "crack" the case of lab-grown eggs. (Just a bit of humor there, we're not egg-actly cracking anything). Imagine a chicken-free egg scramble that tastes, cooks, and even scrambles like the real thing. Mind-boggling, isn't it? Well, with the application of cellular agriculture techniques to avian cells, this isn't just a whimsical fantasy. No yolk about it, we're on the brink of something egg-straordinary.

As for seafood, it's time to set sail into uncharted waters. Our oceans, quite literally, are being fished out, and the cultured seafood industry could just be the lifeboat we need. There's something a bit 'fishy' about lab-grown seafood, you might say? Well, seafood cultivated in bioreactors does not carry the risk of ocean pollutants or the guilt of bycatch. Moreover, a serving of lab-grown fish promises to be as much a burst of Omega 3 as its wild counterpart.

The concept of lab-grown dairy, eggs, and seafood might be a lot to digest. There will be those who turn their noses up at it, while others will relish the chance to savor guilt-free, sustainable food. But before we delve too deep into these choppy waters, it's important to note that these developments are in the early stages. We're not there yet, but oh, what a journey it will be.

And of course, this isn't just about offering substitutes. The technology also opens up a realm of possibilities for entirely new foods that are simply unimaginable within the constraints of traditional agriculture. A fifth type of milk, perhaps? An egg with built-in flavor profiles? A whole new species of seafood, ethically created? Now there's food for thought.

We shouldn't overlook the fact that the success of these lab-grown foods hinges on the same factors we discussed for cultured meat: scientific progress, regulatory approval, consumer acceptance, and market scalability. But it's not just a re-run of the same old show. There are unique challenges and opportunities in each of these food categories, adding another layer of intrigue to this scientific plot.

Is the world ready to embrace this brave new food-scape? It's hard to say. After all, we're not in the prediction business. We're just here, sipping our hypothetical lab-grown milk, nibbling on our imaginary cultured cheese, flipping through the pages of the future in wonder. And maybe, just maybe, that future isn't as far away as we think.

In this march towards the horizon of cellular agriculture, we're not just stepping into the future. We're crafting it, bite by delicious, sustainable bite. Lab-grown meat was just the appetizer. Now it's time for the main course. Bon appétit!

Now, who's ready to discuss the roles of restaurants and chefs in this burgeoning foodscape?

The Role of Restaurants and Chefs

As we journey further into the realm of lab-grown foodstuffs, one might wonder, "What's cooking in the professional kitchens?" Ah, an excellent question and timely, too, as we're about to slice and dice our way through the role of restaurants and chefs in this grand culinary metamorphosis.

When it comes to the restaurant industry, these pioneering establishments could be the proverbial silver platter on which cultured cuisine makes its grand entrance. After all, restaurants are the gastronomic gatekeepers, the curators of our palates. They're the ones who transformed sushi from a peculiar foreign dish to a mainstream craving, and quinoa from a what-now to a staple grain. With cultured meat, they could play an instrumental role in shifting public perception from "Hmm, lab-grown?" to "Mmm, grown in a lab!"

Speaking of chefs, those maestros of flavor and artisans of the edible, imagine what they could do with lab-grown ingredients. They are, after all, known for their alchemical prowess, turning raw, often unappetizing ingredients into dishes that tantalize the senses. Imagine, if you will, a future episode of a popular cooking show. The mystery box opens, and there it is - a slab of cultured Wagyu beef. The chefs would probably take it in stride. A pinch of this, a dash of that, and voila! A masterpiece served with a side of sustainability.

But this isn't merely about chefs creating haute cuisine for the adventurous few. The real magic lies in the potential for mass acceptance. Picture your favorite fast-food chain serving a cultured beef burger or a pizza place offering lab-grown mozzarella. The sight of these familiar foods, albeit with a futuristic twist, could make the notion of cultured cuisine more palatable to the everyday consumer.

And let's not underestimate the power of a celebrity chef endorsement. If a culinary luminary were to embrace cultured cuisine, it could boost public interest and acceptance, much like the way they've ignited food trends in the past (avocado toast, anyone?). So, while lab-grown truffles aren't on the menu just yet, with a few high-profile chefs leading the charge, who knows what could be next?

Restaurants and chefs also play a pivotal role in the narrative of food. They help tell the story of dishes, ingredients, and culinary traditions, often introducing us to flavors and cultures we've never encountered before. In this respect, their influence in shaping the story of lab-grown food cannot be overstated.

There is, however, a soupçon of controversy to consider. Not every chef or restaurant will be ready to jump on the bandwagon. Some might argue that cultured meat lacks the authenticity of traditional farming practices. Others might harbor concerns about the taste, texture, or just the general "yuck" factor. After all, overcoming skepticism is like trying to perfectly caramelize onions—it takes time and just the right amount of heat.

It's also worth considering the practicalities of integrating lab-grown ingredients into professional kitchens. Chefs would need to learn how to properly handle and cook these new products. Would a lab-grown steak grill the same way as a conventional one? Does cultured salmon smoke as well as its wild counterpart? Only time, and a host of brave taste-testers, will tell.

Ultimately, the role of restaurants and chefs in our culinary future could be like that of a seasoned cast iron skillet—integral, versatile, and able to handle the heat. With their creativity, they have the potential to turn this scientific achievement into a gastronomic revolution. They may serve up our first experiences of these new foods, guiding us as we take those initial, tentative bites into a future where our meals are not just delicious, but also sustainable and ethical.

The stage is set, the ingredients are almost ready, and the world is eagerly waiting to see what culinary delights will be plated up. So, let's lift the cloche on this brave new world of cuisine and tuck in. Bon appétit! Now, what could a future dinner plate in 2050 possibly look like?

The Future Plate: Imagining the Dinner of 2050

Let's prop open the doors of our imagination and take a tantalizing trek into the world of 2050, a world where the words "lab-grown" prompt drooling rather than dubiousness. As we perch ourselves at the dinner table of the future, we find ourselves faced with an array of dishes that may seem more at home in a science fiction novel than a dining room.

First course: a starter of oysters, a symphony of the sea on your palate. But these are no ordinary mollusks. Far from the seabed, they were cultured in a laboratory, their development carefully controlled to create the perfect medley of saltiness and creaminess. Piled high on a platter, these pearls of the ocean taste just as luscious as their wild counterparts, yet with one key difference: their existence doesn't disturb a single ocean ecosystem.

Next up, a hearty stew, thick and warming. The meat in this pot, however, never roamed the fields nor graced a barn. Instead, it grew from a few cells in a petri dish to become succulent chunks of beef, perfectly marbled and tender. Without the associated environmental toll or animal welfare issues, each spoonful is savored with a sense of smug satisfaction. Who said you couldn't have your beef stew and eat it too?

And let's not forget the side of vegetables, grown not in soil but in vertical farms. These multi-storied greenhouses stacked like a botanist's version of high-rise buildings, harness the power of controlled environments and hydroponics to maximize yield and reduce water usage.

From farm to table, but without the farm and certainly without the table-long distance.

Speaking of vegetables, look closely, and you'll see something astonishing – those aren't ordinary carrots on your plate. They've been enhanced, engineered to contain higher levels of nutrients. Picture that, a carrot with an impressive CV.

Then there's the grilled salmon with a squeeze of lemon. While it flakes just right, this fish never swam upstream. It's another product of cellular agriculture, produced without the overfishing and the bycatch issues that plague our current seafood industry. This fishy dish is one less drop in the ocean of our ecological problems.

Wine? Of course. Engineered yeast has been put to work, producing vintage-quality wines without the grapes. In vino veritas indeed, especially if the veritas is about decreasing water usage and land pressures.

And for dessert? How about a scoop of ice cream, so creamy and decadent that it makes you think of summer days and the faint chime of an ice cream truck. Only, this isn't derived from a cow. It's produced using engineered yeast that produces real milk proteins, recreating the same creamy delight without needing to nudge a single bovine.

Even the packaging of the future takes center stage. Biodegradable, compostable, and even edible containers mean you can forget about the ever-looming plastic monster. Waste? Not on this menu.

Finally, as you lean back, satiated, you appreciate the check. Not just because it's affordable, but because it's free of the environmental guilt that often comes with our contemporary diets. This is the promise of our culinary future, one where we can feast without the famine of resources.

Looking ahead to the future of food might seem like a flight of fancy. But remember, today's norm was once the extraordinary. So, let's raise a glass (of lab-grown wine, of course) to a future where our dinner plates aren't just about nourishing us, but also about preserving the planet. Here's to the future!

Conclusion: Cultivated Cuisine and its Place in Our World

As we take our last bite from this proverbial meat of discourse, let us step back and examine the plate we've just cleared. The realm of lab-grown meat, or cultivated cuisine as we've affectionately termed it, isn't just about producing patties without a pasture or making mignon without a moo. It's a radical rethinking of what we eat, how we produce it, and our relationship with the world around us.

Cultivated cuisine presents an opportunity to rewire our food system for the better. Imagine an agricultural landscape where we can cut back on deforestation, reduce greenhouse gas emissions, and still produce enough meat to satisfy our cravings. Sounds like a pie-in-the-sky? Well, remember when people thought flying in the sky was a pie-in-the-sky idea, too.

Yet, it's not just about environmental efficiencies. The science of lab-grown meat is quite a marvel, a testament to human ingenuity and our constant striving to better ourselves. Picture a world where we can grow a juicy steak without the need for a whole cow. It's like baking a cake but only needing a single slice to get there. A pinch of cells, a dash of nutrients, a good dose of patience, et voila! – a lab-grown feast.

Our table manners need to change, too. Cultivated cuisine challenges our culinary traditions and tastes, nudging us to embrace the new and unfamiliar. It begs us to ask: what does it mean to eat meat if an animal isn't involved? Can a steak still sizzle if it's born from a bioreactor? Food for thought, indeed.

And what about health and nutrition? Here again, cultivated cuisine shows promising potential. Imagine if we could design meat to be healthier, lower in saturated fats, perhaps even fortified with essential nutrients. Could we create a future where bacon is not a guilty pleasure but a health food? Okay, that might be stretching it, but it's a tasty thought.

Of course, this new plate comes with its own share of hiccups. Technological hurdles, economic viability, regulatory landscapes – they're like the unwanted side dishes we didn't order but got served anyway. Yet, history shows us time and again that we humans have a knack for overcoming obstacles, especially when it involves our stomachs.

The world's dining scene has a role to play as well. Chefs and restaurants, as purveyors of taste and tradition, can help ease lab-grown meat onto our plates and into our palates. Maybe one day, Michelin stars will be awarded not just for culinary artistry but also for culinary responsibility.

But let's not forget the real power players here: you and me. Our perceptions and appetites have the power to shape this nascent industry. Will we accept lab-grown meat as a viable alternative, or will we continue to crave the conventionally farmed counterpart? Only time (and our taste buds) will tell.

On the global chessboard, the play of lab-grown meat could be transformative. It could bring about shifts in agricultural practices, economic dynamics, even geopolitical power. A future where nations spar over bioreactor technologies rather than oil reserves – now that's food for thought.

So as we step away from our dining table of discourse, we carry with us an understanding of how cultivated cuisine might redefine our relationship with food. Will lab-grown meat be the main course of our future meals? Perhaps. But even if it's not, it's started a conversation that we desperately need to have. Because when it comes to our food system, change is not just desirable; it's necessary. Here's to that change. Or as they say in the culinary world, Bon Appétit!

Appendix

Glossary of Terms

Bioreactor: A device or system that supports a biologically active environment, allowing for the culture of cellular products in controlled conditions.

Biotechnology: Technology based on biology, harnessing cellular and biomolecular processes to develop technologies and products that help improve our lives and the health of our planet.

Cellular Agriculture: The field of producing agricultural products from cell cultures rather than whole plants or animals.

Cellular Biology: A branch of biology that studies the different structures and functions of the cell, which is the basic unit of life.

Cultivated Cuisine: A term used in this book to refer to food products, particularly meat, produced through cellular agriculture.

Cultured Meat: Meat produced by culturing animal cells directly, rather than from raising and slaughtering whole animals. Also known as lab-grown meat, cell-based meat, clean meat, etc.

Economic Viability: The capability of developing and surviving as a relatively independent social, economic or political unit. In this context, it refers to the financial feasibility of producing and selling lab-grown meat at a competitive price.

Farmers and Industry: Refers to the traditional agricultural sector and meat industries, which could be affected by the rise of lab-grown meat.

Food Culture: The practices, attitudes, and beliefs surrounding the production, distribution, and consumption of food.

Greenhouse Gas Emissions: Gases in Earth's atmosphere that trap heat, contributing to the greenhouse effect. Major sources include carbon dioxide from burning fossil fuels and from deforestation, and methane from livestock and other agricultural practices.

International Scene: Refers to the global landscape of development, investment, and acceptance of lab-grown meat.

Lab-Grown Dairy, Eggs and Seafood: Other products apart from meat that can potentially be produced via cellular agriculture.

Legal and Regulatory Landscapes: The laws and regulations pertaining to a particular industry or field, in this case, lab-grown meat.

Marketing and Branding Strategies: Techniques and methods used to create and maintain a certain perception of a product or company in customers' minds.

Nutritional Benefits: Advantages gained from consuming foods that contribute to good health and adequate nutrition.

Public Opinion: The collective opinion of a group of individuals on a particular topic or issue, which in this book refers to attitudes towards lab-grown meat.

Restaurants and Chefs: Key players in the food industry who have a significant influence on food trends and consumer acceptance of new food products like lab-grown meat.

Technological Barriers: Obstacles related to technology that prevent or hinder progress in a given field.

The Future Plate: A term used in this book to describe the potential future state of our diets, heavily influenced by the possibilities of cellular agriculture.

About the Author

Gordon Rayner, an avid backyard griller and dedicated family man, brings a unique blend of scientific knowledge and industry experience to the table, which makes him an authoritative voice on the subject of lab-grown meat.

Rayner's journey began with a degree in Biology, stoking an intellectual curiosity that proved instrumental in understanding the intricate world of cellular agriculture. His fascination for science, coupled with his passion for sustainable food practices, led him down the path of exploring the future of food.

Beyond the halls of academia, Rayner honed his practical understanding of the food industry in the trenches of agribusiness. As a marketing executive in the farming industry, he's had his finger firmly on the pulse of the rapidly changing landscape of food production. This dual expertise in both the theory and practice of agriculture has equipped him with a holistic view of the sector - a perspective that resonates throughout his writing.

Yet, Rayner is more than just a biologist turned agribusiness professional. He's a man who knows how to balance the scales of life. When he isn't exploring the future of food or contributing to today's farming industry, Rayner devotes his time to his family, perpetuating his love for food through epic backyard grilling sessions that have become the stuff of local legend.

And there's another twist in Rayner's tale. In the 80s, he was not only navigating the realms of science and food but was also an award-winning roller skater. Rayner ruled the rinks with flair and precision, mirroring the same dedication and passion he now brings to understanding the burgeoning field of lab-grown meat.

With 'Cultivated Cuisine: The Future of Lab-Grown Meat', Gordon Rayner invites you to join him at the intersection of science, industry, and passion, serving up a comprehensive and engaging exploration of the world of cellular agriculture.